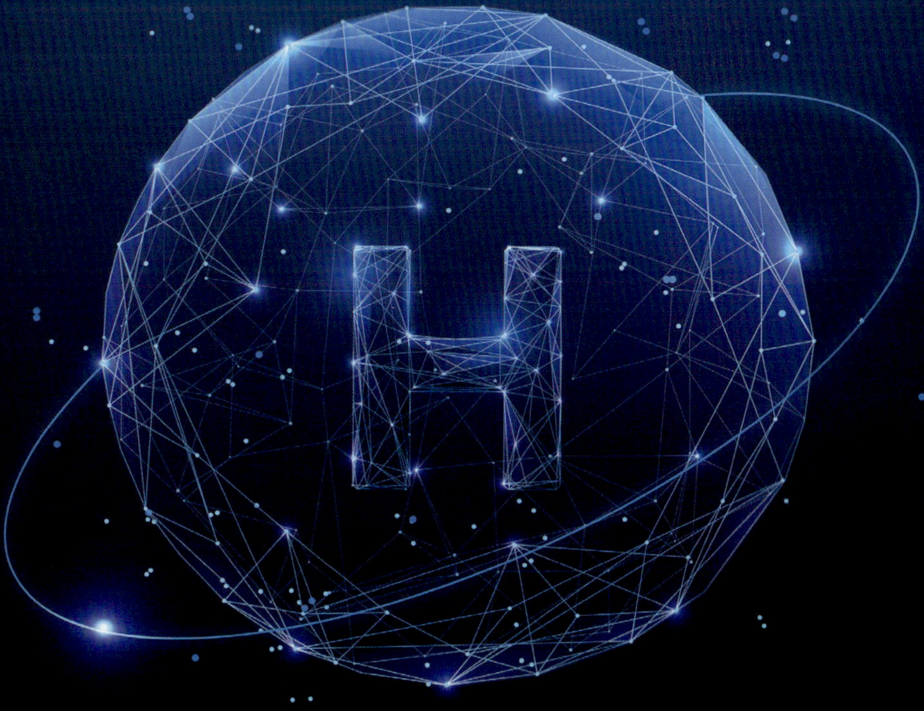

ELEMENTS

BY MARIE ROESSER

Please visit our website, www.enslow.com.
For a free color catalog of all our high-quality books, call toll free 1-800-398-2504 or fax 1-877-980-4454.

Library of Congress Cataloging-in-Publication Data

Names: Roesser, Marie, author.
Title: Elements / Marie Roesser.
Description: New York : Enslow Publishing, [2026] | Series: Chemistry in review | Includes bibliographical references and index.
Identifiers: LCCN 2024036173 (print) | LCCN 2024036174 (ebook) | ISBN 9781978542884 (library binding) | ISBN 9781978542877 (paperback) | ISBN 9781978542891 (ebook)
Subjects: LCSH: Chemical elements–Juvenile literature.
Classification: LCC QD466 .R574 2026 (print) | LCC QD466 (ebook) | DDC 546–dc23/eng/20240905
LC record available at https://lccn.loc.gov/2024036173
LC ebook record available at https://lccn.loc.gov/2024036174

Published in 2026 by
Enslow Publishing
2544 Clinton Street
Buffalo, NY 14224

Portions of this work were originally authored by Kennon O'Mara and published as *Elements* (A Look at Chemistry). All new material in this edition is authored by Marie Roesser.

Designer: Claire Zimmermann
Editor: Therese Shea

Photo credits: Cover, p. 1 (main image) Deemerwha studio/Shutterstock.com; series art (molecule header image) jijomathaidesigners/Shutterstock.com; p. 5 ZikG/Shutterstock.com; p. 9 audioundwerbung/iStock; p. 11 Inkoly/Shutterstock.com; p. 15 (upper) Garfieldbigberm/Shutterstock.com; p. 15 (lower) Cozine/Shutterstock.com; p. 17 Courtesy of Science History Institute; p. 18 Atomic_resolution_Au100.jpg/Wikimedia Commons; p. 19 Frau aus UA/Shutterstock.com; p. 21 Kramskoy_Mendeleev_01.jpg/Wikimedia Commons; p. 21 (inset) 1869-periodic-table.jpg/Wikimedia Commons; p. 23 (periodic table) Peter Hermes Furian/Shutterstock.com; p. 25 PeopleImages.com - Yuri A/Shutterstock.com; p. 27 (top) New Africa/Shutterstock.com; p. 27 (bottom) Miriam Doerr Martin Frommherz/Shutterstock.com; p. 29 Artist_Concept_-_Astronaut_Performs_Tethering_Maneuvers_at_Asteroid.jpg/Wikimedia Commons.

Printed in China

CPSIA compliance information: Batch #QSENS26: For further information contact Enslow Publishing, at 1-800-398-2504.

CONTENTS

Words in the glossary appear in **bold** the first time they are used in the text.

BUILDING BLOCKS

Do you like to build with blocks? In nature, everything that takes up space is made of the building blocks called elements. That's all solids, **liquids**, and gases—everything you can see and can't see. Elements make up everything in the **universe**!

LEARN MORE

An element is matter that can't be broken down into another kind of matter.

There are 118 known elements. Each is made of one kind of atom. An atom is made of smaller **particles** called neutrons, protons, and electrons. If you could look into the atoms of a pure element, you'd see all the atoms have the same number of protons.

ATOMIC STRUCTURE

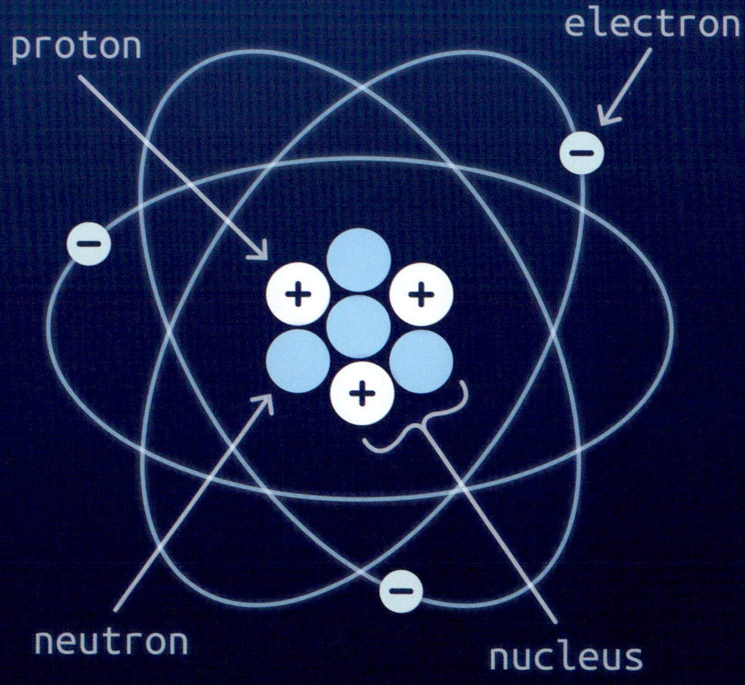

LEARN MORE

Scientists think only 94 of the 118 known elements exist in nature on Earth. The rest of the elements are made in labs.

MOLECULES MAKE MATTER

We can't see a single atom with our eyes. It's tiny! It takes a lot of atoms to make something we can see. Atoms may join to make molecules. Molecules make matter around us. For example, hydrogen is usually made up of joined pairs of hydrogen atoms.

LEARN MORE

Hydrogen is the most **abundant** element, but it's uncommon in a pure form on Earth.

Atoms of different elements can join together too. When atoms of different elements combine, they make a molecule called a compound. The gas carbon dioxide is a compound. Each molecule has one atom of the element carbon and two atoms of the element oxygen.

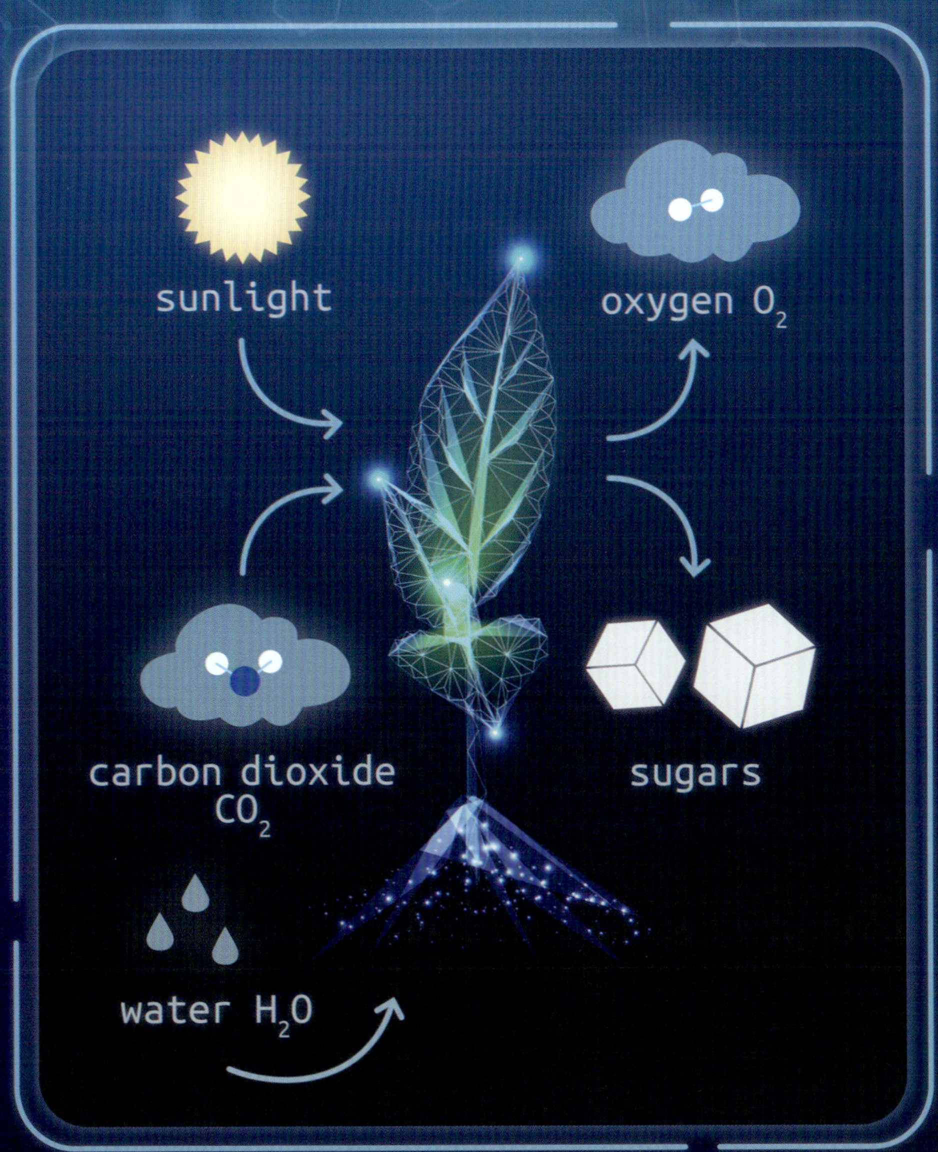

LEARN MORE

Plants use carbon dioxide to make food in a process called photosynthesis.

CHANGING FORMS OF MATTER

All elements can exist as a gas, solid, or liquid. Very cold and very hot **temperatures** can change an element's form. Temperatures affect how molecules act. Bonds between atoms can break or form, causing a change in matter.

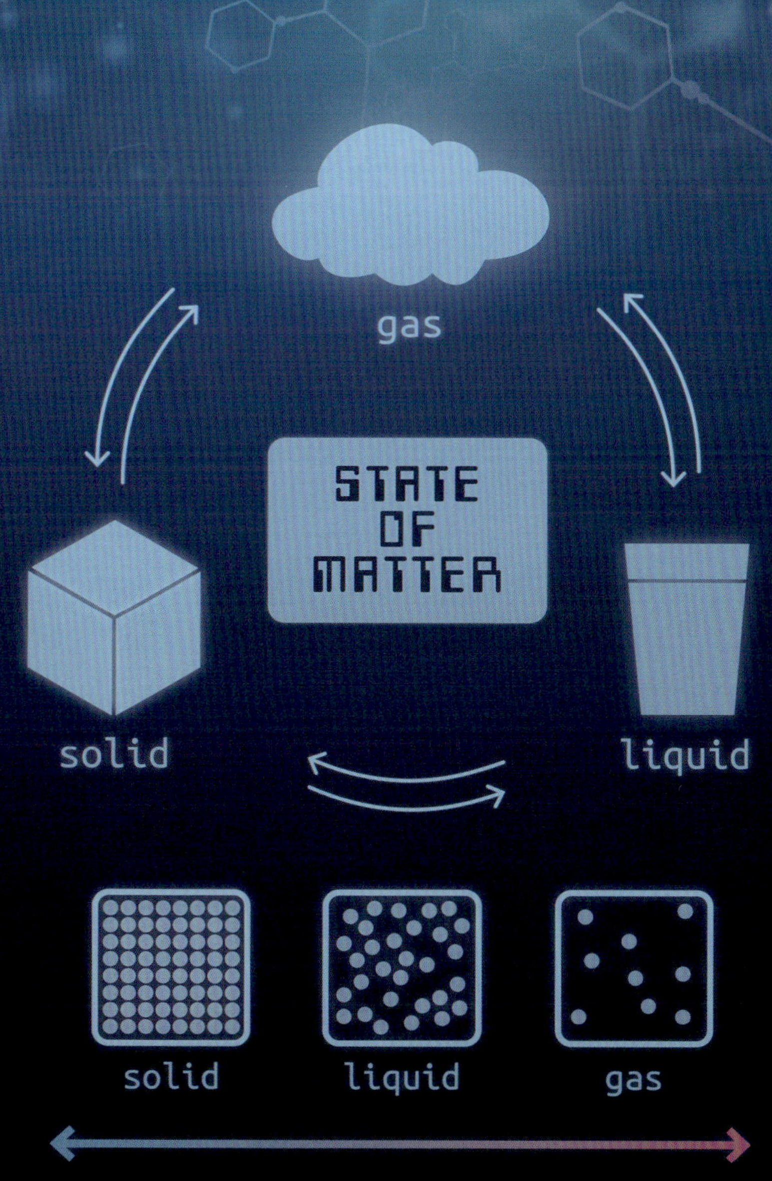

LEARN MORE

Mercury and bromine are the only elements that are liquid at normal room temperature.

Water is a compound. Each molecule has two hydrogen atoms and an oxygen atom. Hydrogen and oxygen are both gases at room temperature. However, when they combine, they go through a **chemical** change and a **physical** change that produces liquid water.

LEARN MORE

When water joins to the compound sodium chloride, salt water forms!

EARLY CHEMISTRY

Long ago, ancient Greek thinkers broke down all matter into four "elements": water, earth, fire, and air. In the 1600s, British-Irish scientist Robert Boyle thought these couldn't be true elements because they didn't form matter.

LEARN MORE

We know water isn't an element because it can be broken down into simpler substances, or matter. Elements can't be broken down further.

Some scientists **identified** elements by burning matter. Others used electricity. As scientists became more skilled in breaking down matter, new elements were discovered. Scientists also began to understand more about atoms. Later, they even saw atoms through special **microscopes**!

atomic view of gold

LEARN MORE

Metals were some of the earlier known elements. These included copper, lead, gold, silver, and iron, which were often found in **ore**.

ORDERING THE ELEMENTS

Scientists learned more about elements' **characteristics**. They realized the elements could be **organized** in a helpful way. Russian scientist Dmitri Mendeleev organized the first periodic table of elements in 1869. He noticed some elements seemed to be missing from the **patterns** he worked out. These elements hadn't been identified yet.

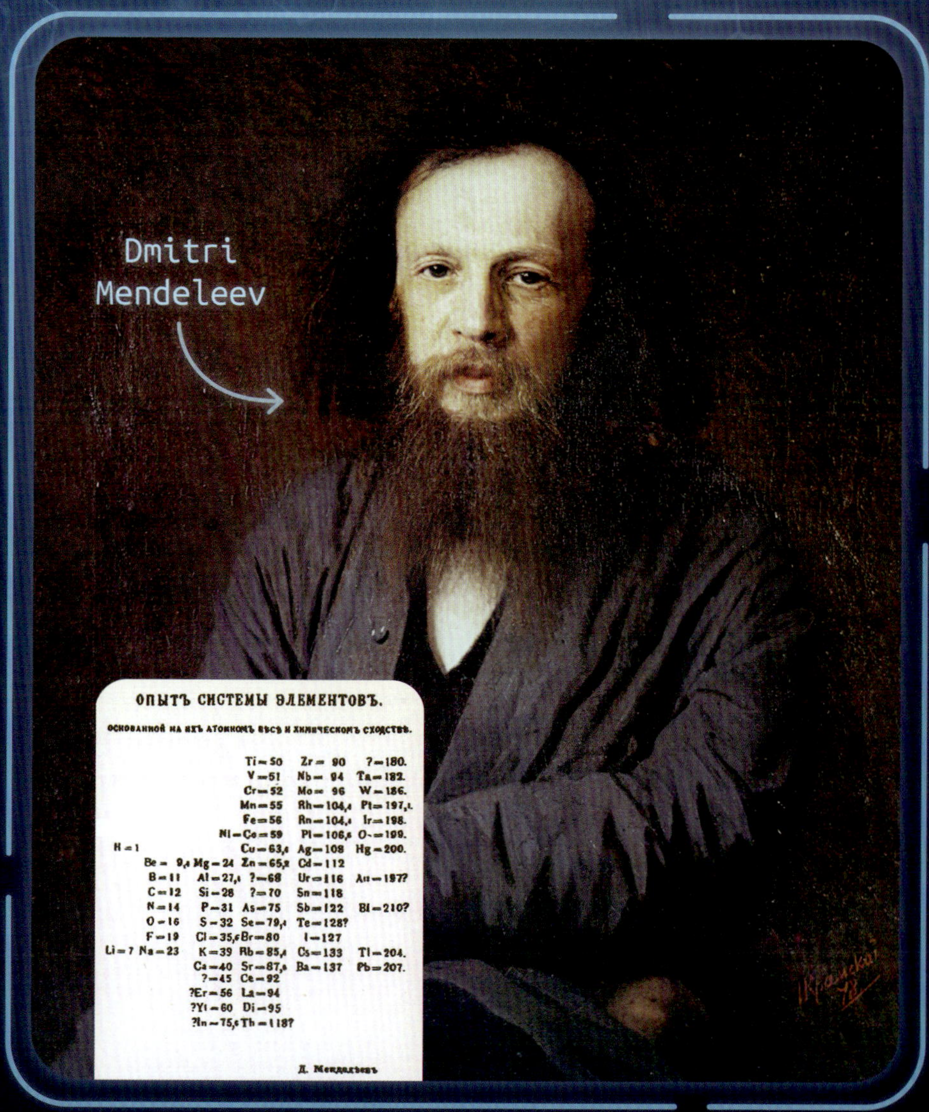

ОПЫТЪ СИСТЕМЫ ЭЛЕМЕНТОВЪ.

ОСНОВАННОЙ НА ИХЪ АТОМНОМЪ ВѢСѢ И ХИМИЧЕСКОМЪ СХОДСТВѢ.

			Ti=50	Zr=90	?=180.
			V=51	Nb=94	Ta=182.
			Cr=52	Mo=96	W=186.
			Mn=55	Rh=104,4	Pt=197,4
			Fe=56	Rn=104,4	Ir=198.
			Ni=Co=59	Pl=106,6	O-=199.
H=1			Cu=63,4	Ag=108	Hg=200.
	Be=9,4	Mg=24	Zn=65,2	Cd=112	
	B=11	Al=27,4	?=68	Ur=116	Au=197?
	C=12	Si=28	?=70	Sn=118	
	N=14	P=31	As=75	Sb=122	Bi=210?
	O=16	S=32	Se=79,4	Te=128?	
	F=19	Cl=35,5	Br=80	I=127	
Li=7	Na=23	K=39	Rb=85,4	Cs=133	Tl=204.
		Ca=40	Sr=87,6	Ba=137	Pb=207.
		?=45	Ce=92		
		?Er=56	La=94		
		?Yt=60	Di=95		
		?In=75,6	Th=118?		

Д. Менделѣевъ

LEARN MORE

Early **versions** of the periodic table, like many today, shorten each element's name to one or two letters. This is its atomic symbol. For example, "H" is for hydrogen.

In today's periodic tables, an element's place in its row, or period, is according to its atomic number, or the number of protons in a single atom. An atom's atomic weight may also be listed. This is about equal to the number of protons added to the number of neutrons.

PERIODIC TABLE

atomic number → 1
H ← symbol
name → Hydrogen

1 H Hydrogen																	2 He Helium
3 Li Lithium	4 Be Beryllium											5 B Boron	6 C Carbon	7 N Nitrogen	8 O Oxygen	9 F Fluorine	10 Ne Neon
11 Na Sodium	12 Mg Magnesium											13 Al Aluminium	14 Si Silicon	15 P Phosphorus	16 S Sulfur	17 Cl Chlorine	18 Ar Argon
19 K Potassium	20 Ca Calcium	21 Sc Scandium	22 Ti Titanium	23 V Vanadium	24 Cr Chromium	25 Mn Manganese	26 Fe Iron	27 Co Cobalt	28 Ni Nickel	29 Cu Copper	30 Zn Zinc	31 Ga Gallium	32 Ge Germanium	33 As Arsenic	34 Se Selenium	35 Br Bromine	36 Kr Krypton
37 Rb Rubidium	38 Sr Strontium	39 Y Yttrium	40 Zr Zirconium	41 Nb Niobium	42 Mo Molybdenum	43 Tc Technetium	44 Ru Ruthenium	45 Rh Rhodium	46 Pd Palladium	47 Ag Silver	48 Cd Cadmium	49 In Indium	50 Sn Tin	51 Sb Antimony	52 Te Tellurium	53 I Iodine	54 Xe Xenon
55 Cs Caesium	56 Ba Barium	57 La* Lanthanum	72 Hf Hafnium	73 Ta Tantalum	74 W Tungsten	75 Re Rhenium	76 Os Osmium	77 Ir Iridium	78 Pt Platinum	79 Au Gold	80 Hg Mercury	81 Tl Thallium	82 Pb Lead	83 Bi Bismuth	84 Po Polonium	85 At Astatine	86 Rn Radon
87 Fr Francium	88 Ra Radium	89 Ac** Actinium	104 Rf Rutherfordium	105 Db Dubnium	106 Sg Seaborgium	107 Bh Bohrium	108 Hs Hassium	109 Mt Meitnerium	110 Ds Darmstadtium	111 Rg Roentgenium	112 Cn Copernicium	113 Nh Nihonium	114 Fl Flerovium	115 Mc Moscovium	116 Lv Livermorium	117 Ts Tennessine	118 Og Oganesson

*	58 Ce Cerium	59 Pr Praseodymium	60 Nd Neodymium	61 Pm Promethium	62 Sm Samarium	63 Eu Europium	64 Gd Gadolinium	65 Tb Terbium	66 Dy Dysprosium	67 Ho Holmium	68 Er Erbium	69 Tm Thulium	70 Yb Ytterbium	71 Lu Lutetium
**	90 Th Thorium	91 Pa Protactinium	92 U Uranium	93 Np Neptunium	94 Pu Plutonium	95 Am Americium	96 Cm Curium	97 Bk Berkelium	98 Cf Californium	99 Es Einsteinium	100 Fm Fermium	101 Md Mendelevium	102 No Nobelium	103 Lr Lawrencium

LEARN MORE

The columns of the periodic table form groups that have similar chemical characteristics. For example, certain gases are in one column and certain metals are in another.

ELEMENTS ARE IN YOU!

The first 18 elements in the periodic table—the elements with atomic numbers 1 through 18—make up most of the matter in the universe. About 96 percent of our body mass is made of four elements: oxygen, carbon, hydrogen, and nitrogen.

LEARN MORE

The most abundant element in our bodies is oxygen, which makes up around 65 percent.

ALLOTROPES

An element can exist in different forms in nature. For example, atoms of the element carbon may bond in different ways to take the form of graphite, diamond, or coal. These different substances that a single element can make are called allotropes.

LEARN MORE

The graphite in your pencil is made of the same element as a diamond!

STILL LOOKING

Scientists are trying to create elements 119 and 120 in labs. They do this by adding a proton or neutron to another element. It's not easy to do. New elements decay, or break down, really fast too. New elements may be found in space someday!

LEARN MORE

Some **asteroids** may have heavier elements, beyond those in the periodic table. Scientists are excited to explore this possibility someday!

ELEMENTS OF INTEREST

Element	Atomic Number	Symbol	Fun Fact
helium	2	He	atoms so light they can escape Earth's gravity
lithium	3	Li	metal light enough to float on water
neon	10	Ne	gas mainly found in stars
aluminum	13	Al	most abundant of Earth's metals
potassium	19	K	needed for animal and plant life
bromine	35	Br	named for Greek word for "stench" due to its bad smell
mercury	80	Hg	metal that is liquid at room temperature
curium	96	Cm	glows in the dark

GLOSSARY

abundant: Present in great amounts.

asteroid: A small rocky body that goes around the sun.

characteristic: A feature that something has that helps identify it.

chemical: Having to do with atomic or molecular changes.

identify: To find out names or features.

liquid: Matter that can flow freely.

microscope: A tool to view very small objects so they can be seen much larger and more clearly.

organize: To put together in an orderly way.

ore: Matter in the ground from which a valuable metal can be removed.

particle: A very small piece of something.

pattern: Something that happens or is done in a repeated and regular way.

physical: Having to do with a form you can touch or see.

temperature: How hot or cold something is.

universe: Everything that exists.

version: A different form of something.

FOR MORE INFORMATION

BOOKS

Chad, Jon. *The Periodic Table of Elements: Understanding the Building Blocks of Everything.* New York, NY: First Second, 2023.

Dingle, Adrian. *My Book of the Elements: A Fact-Filled Guide to the Periodic Table.* New York, NY: DK Publishing, 2024.

WEBSITE

The Ring of Truth: Two Hydrogen Atoms & One Oxygen Atom
thekidshouldseethis.com/post/the-ring-of-truth-two-hydrogen-atoms-one-oxygen-atom
Watch water form from hydrogen and oxygen!

INDEX